ARROYO CENTER

T0108954

Army Stock Positioning

How Can Distribution Performance Be Improved?

Adam C. Resnick, Jeremy M. Eckhause, James Syme

Prepared for the United States Army

For more information on this publication, visit www.rand.org/t/RR1375

Library of Congress Cataloging-in-Publication Data is available for this publication.
ISBN: 978-0-8330-9643-2

Published by the RAND Corporation, Santa Monica, Calif.
© Copyright 2017 RAND Corporation
RAND® is a registered trademark.

Support RAND

Make a tax-deductible charitable contribution at
www.rand.org/giving/contribute

www.rand.org

Preface

The position of inventoried stock within the U.S. Army's logistical network greatly influences its operational success. The ability of the Army to deliver materiel as needed on a timely basis to support training and maintain readiness can ultimately decide mission outcome. Improving delivery of materiel can improve overall performance.

This report, produced under a project titled "Stock Positioning," examines how the Army can improve distribution performance by decreasing customer wait times and reducing costs. The research focuses on how the Army can improve delivery of a specific class of items—secondary items weighing at least 50 pounds—that comprise a relatively small number of total items the Army delivers but account for a great deal of the weight of items it delivers. It focuses on how the Army can improve its distribution system through strategic prepositioning and adjustments to source preferences.

This report will be of direct interest to Army Materiel Command (AMC) G-3/4, charged with leading the Army's resource-management processes. However, the opportunities identified in this research can be applied broadly across AMC-managed items, and integrated into future business processes.

This research was sponsored by Army Materiel Command G-3/4 and conducted within the RAND Arroyo Center's Forces and Logistics Program. RAND Arroyo Center, part of the RAND Corporation, is a federally funded research and development center sponsored by the United States Army.

The Project Unique Identification Code (PUIC) for the project that produced this document is HQD156892.

Contents

Figures and Tables

Figures

Tables

Summary

The position of inventoried stock within the U.S. Army's logistical network is inseparably linked to operational success. The ability of the Army to deliver materiel as needed on a timely basis to support training and maintain readiness can ultimately decide mission outcome. Put another way, should the Army be able to improve delivery of materiel, it will improve its overall performance.

The Army has undergone a number of improvements to its stock-positioning and delivery processes in recent years, and current stock-positioning processes have been effective. Yet, several major changes in recent years have prompted a need to reevaluate how well these processes address the rapidly changing security environment and fiscal constraints. To help the Army continue streamlining its logistics processes, the Army Materiel Command (AMC) G-3/4 tasked the RAND Corporation's Arroyo Center with evaluating current stock-positioning processes and recommending ways to improve responsiveness and reduce associated costs. That is, AMC asked RAND to find ways to improve distribution performance.

This work focuses on a class of items that can be distributed most efficiently by leveraging the capabilities of the Defense Logistics Agency (DLA) distribution network. In particular, we look at heavy secondary items that weigh at least 50 pounds (eliminating end items, such as vehicles, that are shipped by other dedicated modes that would not be consolidated into the DLA network). Altogether, items weighing more than 50 pounds numbered roughly 400,000 of the 4.2 million secondary items in the Army-managed inventory shipped from DLA depots from January 2014 to May 2015 and accounted for 81 percent of the weight of such shipments.

We begin by examining how the Army currently positions its inventoried items. Next, we review options for delivering these items to customers, including the DLA network. We then discuss results of a case study for improving delivery of one particular item: wheel assemblies. We conclude with a summary of our findings.

Current Positioning Processes

The Army has two main sources of inventoried items for distribution: Those it procures new, and those it repairs and redistributes. Delivery location of new items may be determined by contract (possibly constraining options of distribution managers), a topic we will discuss further below.

Items that are repaired at Army depots are, by default, placed in inventory at colocated DLA depots. Item managers can adjust this placement process by stipulating—at the time repairs are ordered—that items should be stored in a different warehouse. Item managers can also initiate orders to move items, most efficiently in truckloads, from one DLA depot to another. Items are available to issue worldwide once they are in inventory at the colocated DLA depot, and frequently they remain there until they are issued.

Item managers may also rely on the National Maintenance Program (NMP) for items. Through the NMP, AMC assigns maintenance for reparable items to repair sites. The NMP assesses the national-level requirement for maintenance workload and enters into agreements with approved sources of repair to perform the work. By managing maintenance centrally, the NMP ensures that work will be performed to the highest standard and reduced the redundant capacity that would exist across Army posts were the work performed locally. NMP repairs may be performed at active- or reserve-component Army installations across the country. When items come out of repair, they are available for issue. As items are continuously being repaired, this process allows AMC to strategically position the finished items to optimize distribution performance. Logistics Readiness Centers that participate in the NMP must store items repaired at their locations until they are sent elsewhere. Alternatively, item managers may ship full truckloads of items to a DLA distribution center. Because item managers are rated by their ability to avoid back orders, they may choose to leave items repaired at NMP sites in place so that they are available to issue to customers directly. This may not, however, boost performance of the overall distribution system.

Strategically positioning items at DLA depots to improve distribution performance is an intuitive but complex decision that can be supported with available data. Such positioning may be accomplished either through procurement of new items that are sent directly to DLA hubs or by shipment of truckloads of items to such hubs for later shipment to ultimate customers. In the next section, we describe the DLA hub system and how it can help improve distribution-system performance for Army-managed inventory.

The DLA Three-Hub Network

The DLA maintains a contiguous United States (CONUS) network of three distribution hubs at San Joaquin, California (DDJC), Red River, Texas (DDRT), and Susquehanna, Pennsylvania (DDSP). These hubs are served by a network of scheduled-truck service delivering items from the hubs to installations within each hub's region at least once and typically several times weekly. For items available both at the hubs and at maintenance centers, delivery by scheduled-truck service from the hubs to installations is often quicker than delivery by commercial truckload or less than truckload (LTL) to installations. Figure S.1 shows the average requisition wait time for Army customers for items delivered by scheduled truck from a DLA hub or by LTL/truckload from a maintenance center, as well as the difference between them. In nearly all cases, delivery from a DLA hub by scheduled-truck service is quicker than LTL/truckload shipping from a maintenance center. In some cases, the difference exceeds five days.

Figure S.1
Average Requisition Wait Time for Army Customers Receiving Items by Scheduled-Truck or LTL/Truckload Service,
January 2014–July 2015

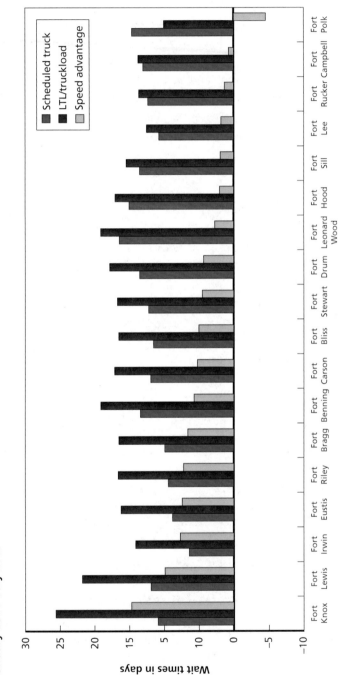

RAND *RR1375-S.1*

DLA established its scheduled-truck network in coordination with the military services to garner savings on shipping costs and improve responsiveness. Delivery of items by scheduled-truck service saves costs to the U.S. Department of Defense (DoD). DLA directly bears the cost of issuing items from its warehouses to customers, recouping these costs through the cost-recovery rate it imposes on items that are sold from the Defense Working Capital Fund to customers from the military services, and from net landed costs it assesses to items received, stored, and issued from its warehouses. The cost savings from distribution efficiencies are then passed along to the military services.

Consider, for example, a 250-pound item that is shipped via LTL directly from a depot to a customer. At $0.60 per pound—and our research found that shipping costs for such items can range from $0.25 to $0.75 per pound—this item will cost $150 to ship. By contrast, were the Army to ship truckloads of such items to a DLA hub, from which they would ship to customers on the DLA scheduled-truck network, then this item would cost only $0.20 per pound to ship for the transshipment leg, with costs for receiving and issuing from DLA, bringing the ultimate costs for shipping this item to $81.

Moving more shipments to the DLA scheduled-truck network can offer substantial savings. Of 400,000 items weighing at least 50 pounds shipped to customers from January 2014 to May 2015, approximately 180,000 were issued to customers who were on a scheduled-truck route. Only about half of these items, however, were delivered by scheduled-truck service; the other half were issued from more-distant DLA depots than the nearest hub and shipped by other modes. In other words, about 90,000 items that could have been delivered by scheduled-truck service were shipped by other modes and from more-distant DLA depots. Had these items been positioned at a DLA hub and shipped on the scheduled-truck service, the Army would have realized more than $13 million in shipping costs, or $800,000 per month in the time period we considered.

The data in Figure S.2 show the projected decrease in requisition wait time (RWT) and savings in distribution costs per month that can be garnered by supporting AMC item managers to position stock proportionate to demand. In current practice, RWT is approximately 16

Figure S.2
Savings in Requisition Wait Time and Distribution Costs Because of Stock Positioning; Army-Manaaged Items Issued to Scheduled-Truck Customers from Distant DLA Hubs

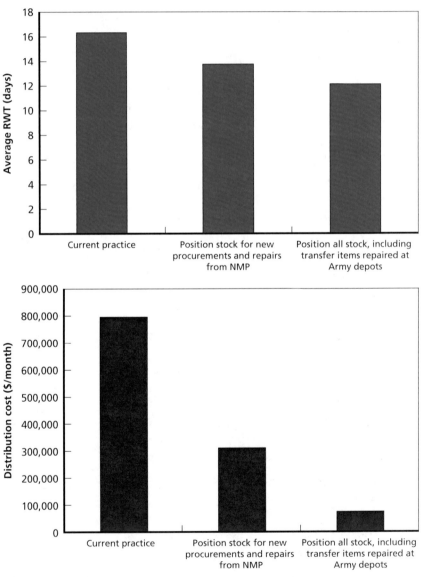

days for items issued from DLA depots by LTL to customers who are on scheduled-truck networks. The estimated shipping costs incurred for Army-managed items (AMI) greater than 50 pounds is approximately $800,000 per month. By positioning new procurements and items coming out of repair at NMP sites, AMC item managers can affect 61 percent of these items. By positioning the balance of items, such as customer returns to DLA depots, and by transferring AMI coming out of maintenance at Army depots when it is advantageous, AMC item managers can generate a decrease in average RWT of more than four days and a cost savings of nearly $800,000 per month.

By strategically considering placement of items before shipment, item managers can improve responsiveness and save costs. If AMC leverages the capabilities of its enterprise resource planning system, the Logistics Modernization Program (LMP) to practice strategic stock positioning, it would change its management principles to differ from current practice in ways that are summarized in Table S.1.

In the following section, a case study of wheel assemblies helps demonstrate how this might work.

Table S.1
Management Principles Under Current Practice and Strategic Stock Positioning

Current Practice	Strategic Stock Positioning
No visibility of regional demand	Transaction code in LMP calculates regional demand
No decision support to position stock	Transaction code in LMP compares demand projection with inventory projection
Item managers place new procurements orders to single warehouse	Item managers can align orders regionally, improving RWT and lowering distribution costs
Repair sites must contact item managers to move items, leads to excess and misaligned inventory	Item managers have decision support to tell repair sites in advance where they should ship completed items
Item managers do not have a business case to move items that come out of repair at Army depots	Item managers can justify moving items to leverage scheduled-truck network, decrease RWT

Improved Positioning of Wheel Assemblies

Wheel assemblies are one of the most-common AMI items repaired. The West Virginia Army National Guard (WV ARNG) facility in Charleston, West Virginia, repairs about 60 percent of wheel assemblies for the Army—nearly 13,000 in 2012 and 2013. In 2012, it shipped all its repaired wheel assemblies to DDRT, which was sensible for meeting demand during the conflicts in Iraq and Afghanistan, but not the most efficient option for meeting demand within CONUS.[1]

To better balance positioning with demand, RAND researchers in March 2013 suggested that the WV ARNG send some wheel assemblies it ships to DDRT to DDSP instead. Figure S.3 illustrates this change.

This change enabled Army customers on scheduled-truck routes from DDSP to receive wheel assemblies via this mode in 2013. Figure S.4 shows the proportion of DLA wheel assembly issues from each DLA hub to customers in the eastern region of the United States from 2012 to 2014—and the increasing proportion that came from the DDSP, the DLA hub in the region, for these customers.

In terms of the same metrics used to project savings to the Army by implementing stock-positioning policies for all AMI, we estimate the effect of change in wheel assembly management. In 2014, the national maintenance program at WV ARNG sent approximately 4,200 wheel assemblies to DDSP for later distribution to customers and 500 wheel assemblies to DDRT in 2014. The estimated cost savings because of wheel assembly positioning are shown in Figure S.5. We estimate that distribution costs decreased by approximately $150,000, and RWT was reduced by one day.

To further improve delivery-system performance, RAND researchers suggested changes in the source-preference logic for filling requisitions in the LMP so that smaller customers not served by

---

[1] Item managers reported that, during the height of operations in Iraq and Afghanistan, DDSP experienced delays distributing materiel because of overwhelming demands placed on it as the DoD primary distribution site on the East Coast. Item managers found they could improve distribution speed by sending items to DDRT for further distribution through the Gulf of Mexico and on to Iraq and Afghanistan via surface freight.

Figure S.3
WV ARNG Wheel Assembly Issues, February 2012–August 2014

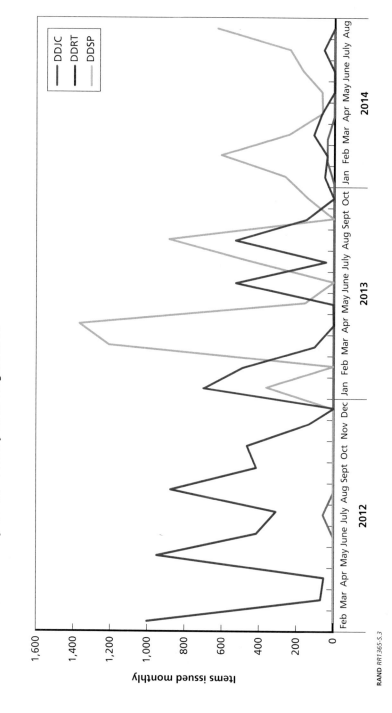

RAND RR1365-S.3

Figure S.4
Wheel Assembly Issues from DLA Hubs to Eastern Region Customers

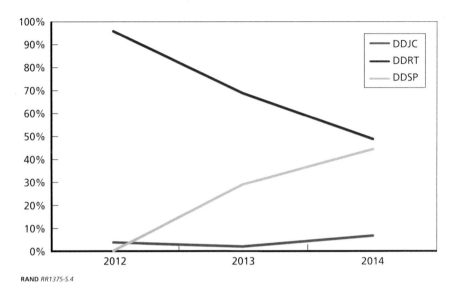

RAND RR1375-S.4

DLA scheduled trucks would be issued items preferentially by the WV ARNG rather than an eastern U.S. region DLA depot. Two of the largest nonscheduled-truck route customers in the eastern U.S. region are the Tennessee and South Carolina ARNG units. The many locations of ARNG units in these states order items independently, but their requests are routed through the central U.S. Property and Fiscal Officer site in each state. On average, the ARNG units in Tennessee and South Carolina are issued 20 to 25 items per month directly from the WV ARNG, but during two months of the pilot change in the source-preference logic, they received 250 items from the WV ARNG and zero items of these types from DLA depots. This, in turn, permitted DLA to issue more items to customers on its scheduled-truck routes.

Conclusion

AMC can improve distribution through stock positioning. It has several opportunities to direct where inventory is located, including when

Figure S.5
Estimated Effect of Wheel Assembly Positioning Policies, WV ARNG Issues to DLA; Reduced Customer RWT and Distribution Costs in 2014

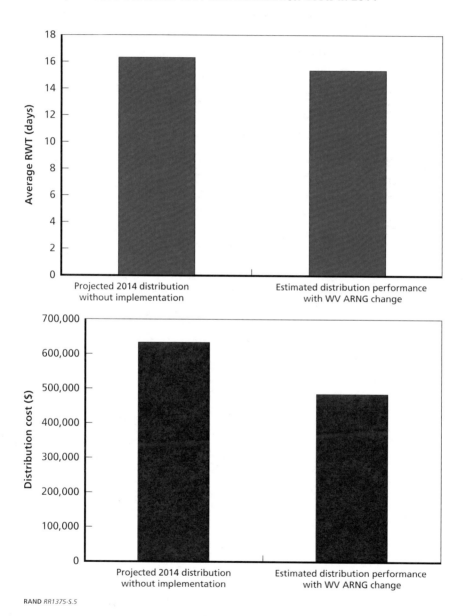

items are newly procured and when they come out of maintenance. It can also adjust its source-preference programming so that customers who should get materiel from DLA hubs and on scheduled-truck deliveries do so, and those who should get it elsewhere do so. Balancing inventory against regional demand can help item managers save costs and reduce requisition wait time.

Acknowledgments

We would like to thank our sponsor, James C. Dwyer, Deputy Chief of Staff, G-3/4, Army Materiel Command, for his project guidance; our action officers, Ron Mailhiot and Stacey Holden; and representatives from Army Materiel Command, who shared their expertise on military logistics and knowledge of the Logistics Modernization Program and provided us with access to data that enabled this project to be a success.

At RAND, we wish to thank Kenneth Girardini, Elvira Loredo, and Bruce Held for their insightful comments and assistance.

Abbreviations

AMC	Army Materiel Command
AMI	Army-managed items
ARNG	Army National Guard
CONUS	contiguous United States
DDAA	Defense Logistics Agency Distribution Anniston, Alabama
DDCT	Defense Logistics Agency Distribution at Corpus Christi, Texas
DDJC	Defense Logistics Agency Distribution at San Joaquin, California
DDRT	Defense Logistics Agency Distribution at Red River, Texas
DDSP	Defense Logistics Agency Distribution at Susquehanna, Pennsylvania
DDTP	Defense Logistics Agency Distribution at Tobyhanna, Pennsylvania
DLA	Defense Logistics Agency
DoD	U.S. Department of Defense
DODAAC	Department of Defense Activity Address Code
ERP	enterprise resource planning
FY	fiscal year
GCSS-A	Global Combat Support System–Army

LCMC	Life Cycle Management Command
LMP	Logistics Modernization Program
LRC	Logistics Readiness Centers
LTL	less than truckload
NMP	National Maintenance Program
OCONUS	outside the contiguous United States
RIC	Routing Identification Code
RWT	requisition wait time
TACOM	Tank-Automotive and Armaments Command (formerly)
USPFO	U.S. Property and Fiscal Officer
WV ARNG	West Virginia Army National Guard

CHAPTER ONE

Introduction

The position of inventoried stock within the U.S. Army's logistical network is inseparably linked to operational success. Materiel is needed on a timely basis to support training and maintain readiness. The ability to fulfill requirements and deliver on time can ultimately decide mission outcome.

Currently, the Army stores Army-managed items (AMI) centrally in warehouses for distribution to Army customers as well to those of the other services as needed. This includes a wide range of inventoried items, from construction materials to vehicle-repair parts. In the early 1990s, the Army began to more strategically position items in the supply chain, refining its inventory and delivery practices by adopting standard, data-driven approaches to materiel replenishment and transfer (Dumond et al., 2001).

Current stock-positioning processes have been effective (Saliba, 2013). Yet, several major changes in recent years have prompted the need to reevaluate how well these processes address the rapidly changing security environment and fiscal constraints. There is also a need to assess new information systems for effectiveness and efficiency. The Army has adopted enterprise resource planning (ERP) software, such as the Logistics Modernization Program (LMP) and the Global Combat Support System–Army (GCSS-A). The Army's Class IX distribution network has also evolved significantly, especially the way it relies on Defense Logistics Agency (DLA) support. The DLA has committed to a three-hub distribution network in the contiguous United

States (CONUS), a move that closely parallels the Army's area-oriented depot strategy.

To help the Army continue streamlining its logistics processes, Army Materiel Command (AMC) G-3/4 asked the RAND Corporation's Arroyo Center to evaluate current stock-positioning processes and recommend ways to improve responsiveness and reduce associated costs—that is, ways to improve *distribution performance*.

How Army Stock Currently Positioned and Study Focus

Figure 1.1 shows the sequence of processes to position AMI. This model approximates both the process for managing items repaired through the National Maintenance Program (NMP) and at Army repair depots, such as Anniston Army Depot and Corpus Christi Army Depot.

The process begins with item managers assessing the future need for materiel. For reparables, item managers perform an annual review to determine how the Army will generate the supply of serviceable items to meet its needs, whether by maintenance of existing items or procurement of new ones. Item managers pass the requirement for items to be repaired to the sources of repair. For AMI that are maintained in the NMP, item managers give the number of each type of item to be repaired in the fiscal year (FY) to NMP representatives. The NMP receives submissions from approved Army repair sites to perform maintenance for nationally managed items. The repair sites submit to AMC the quantity and the cost of work they can perform. The NMP allots workload to repair sites to fulfill the annual requirement. So that the repair sites have a sufficient number of carcasses to work on, AMC and the NMP set parameters in the Standard Army Retail Supply System and GCSS-A that direct units where to send unserviceable items. After NMP sites complete work and generate serviceable items, the items are available for issue in the LMP. At this stage, item managers can decide where to position stock.

For items that are repaired by Army depots or arsenals, the process is similar. Item managers will estimate how many unserviceable items will be available for repair, as well as the projected demand for items. Item managers determine how many items are to be repaired at depots based on these.

After items complete repair at Army depots, they are placed into inventory at the colocated DLA storage depots. Frequently repair and storage depots are so close that repaired items can be transported from a repair site on a forklift and directly deposited in the storage depot. It is common Army practice to store repaired items at colocated depots, although that may not be the location for optimizing distribution performance. In cases where it is beneficial to move some items to a depot to leverage the DLA scheduled-truck network to improve distribution performance, item managers must initiate a stock-transfer order.

Item managers also perform monthly reviews in which they recalculate the projected demand for items based on updated information. Item managers project two types of demand: independent and dependent. They project independent demand based on historical customer demands. This projection is made off a rolling 12-month demand history that projects similar demands in the subsequent 12 months. When item managers have specific knowledge of upcoming customer demands, such as planned deployments or changes in the size and usage of Army vehicle fleets, they can adjust projections for independent demands accordingly.

They also project dependent demands based on scheduled consumption from such activities as repair programs that will consume the items, planned procurement by services other than the Army, and foreign military sales. Based on current inventory of serviceable items and projected demand, item managers determine whether they should initiate repairs or new procurements to generate more serviceable items.

In LMP, items are designated with lead times based on historical times for contracting and supplier response. If a stock outage is projected—that is, if there are fewer items in inventory than needed to meet demands over the item lead time—then item managers will seek new inventory. If there are unserviceable items in inventory that can be used in a repair program, and there is capacity to repair them either through the NMP or at Army depots, then item managers will work to increase the planned repairs for items. If increasing the number of repaired items is not possible, then item managers will seek to procure new items. There is no standard way to determine batch sizes for repair or new procurement items. Batch sizes can represent three, six, nine, or 12 months of demands, based on projected lead times, the quantity at which items can be supplied at a discount, or another quantity determined to make the best use of available funds.

Item managers can greatly affect distribution performance by designating where items should be delivered. They lack, however, a tool in LMP to help them make this decision.

Item managers can mitigate inventory imbalances by checking stock position relative to regional demand on a monthly basis at the same time when they total inventory levels against global demand—which is their current practice. By doing so, they can direct new procurement locations, and position items coming out of maintenance, to prevent large imbalances from occurring.

Analysis Methods, Goals, and Limitations

For this study, we gathered data from Army's Integrated Logistics Analysis Program and from the DLA's Distribution Standard System. To understand how items in the NMP are managed, we interviewed item managers in U.S. Army TACOM (formerly Tank-Automotive and Armaments Command and currently referred to as TACOM Life Cycle Management Command [LCMC]) and other staff in AMC. To document how LMP functions, we interviewed staff at AMC, including staff working in the supply-chain functions at each of the LCMCs.

This report surveys current practice in how stock is positioned by AMC item managers. We describe the business processes that govern these activities and identify where changes may improve distribution performance. We then show a case study of how AMC manages wheel-assembly maintenance and the positive effect of changes AMC made in positioning wheel assemblies after repair. We leave for future research a complete systemwide analysis to project the scale of benefits available through continuing improvement implementations.

Report Organization

In Chapter Two, we look more closely at how AMC positions items after repair and find opportunities for item managers to improve stock positioning, looking at a particular class of items that may benefit most from shipping through DLA's hub network. In Chapter Three, we review the three-hub network that DLA has implemented in CONUS and how AMC can leverage it to improve distribution performance. In Chapter Four, we offer a case study on management of wheel assemblies in the NMP. In Chapter Five, we summarize implications of our research, as well as our recommendations and next steps.

CHAPTER TWO

Process to Position Items

This report focuses on a class of items that can be distributed most efficiently by leveraging the capabilities of the DLA distribution network. These are heavy, bulky items that can be consolidated into truckload shipments to garner lower shipping rates. Such items, when issued to customers by trucks scheduled to run from depots at least weekly, may be distributed to customers more quickly than if they were transported individually by commercial carriers in either truckload or less-than-truckload (LTL) shipping. Items that benefit the most from using lower cost-per-pound shipping rates are those that weigh more than 50 pounds but are not end items, such as vehicles, which are shipped by other dedicated modes that would not be consolidated into the DLA scheduled-truck network. Hence, our analysis focuses on such items.

This chapter addresses the process to position serviceable items made available to customers in several ways: Items that are newly procured, items that come out of maintenance from Army depots, and items that come out of maintenance from NMP sites. The locations of maintenance sites and DLA depots are indicated in Figure 2.1.

There are three main DLA distribution hubs in the United States. DLA Distribution at San Joaquin, California (DDJC), DLA Distribution at Red River, Texas (DDRT), and DLA Distribution at Susquehanna, Pennsylvania (DDSP), are aligned to serve the western, central, and eastern region of the United States. It is from these three depots that the scheduled-truck routes emanate. There are four main Army repair depots that repair items considered in this analysis: Corpus Christi Army Depot, Anniston Army Depot, Tobyhanna Army Depot,

Figure 2.1
Locations of Repair Sites and DLA Depots

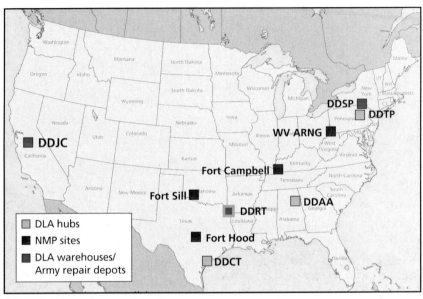

RAND RR1375-2.1

and Red River Army Depot. They are collocated with DLA depots: DLA Distribution at Corpus Christi, Texas (DDCT), DLA Distribution at Anniston, Alabama (DDAA), DLA distribution at Tobyhanna, Pennsylvania (DDTP), and DDRT. DDRT serves a dual purpose as a support for the collocated Army repair depot and as a distribution hub for the central region. The four largest NMP sites by volume are West Virginia Army National Guard (WV ARNG), Fort Hood, Fort Campbell, and Fort Sill.

This chapter further discusses the options for positioning items. We then discuss the opportunities managers have for improving stock positioning.

New Procurement

Many of the serviceable items received at DLA depots come from new procurement. Figure 2.2 shows that 50 percent of the items come from new procurement. When AMC item managers initiate purchase contracts with suppliers, they list one or more DLA depots as delivery locations. Often, for the class of items in this analysis, DDRT is listed as the sole delivery location. It would be very valuable to aid and improve item managers' abilities to order shipments from suppliers to support strategic stock positioning. This can be done when item managers direct contract specialists to specify multiple delivery locations when entering long-term purchase contracts with suppliers. Then, when item managers place purchase orders with suppliers, they have the option to specify the quantity of items that should be delivered to each location.

Figure 2.2
Receipts of Serviceable Items at Highest-Volume Army Depots in United States

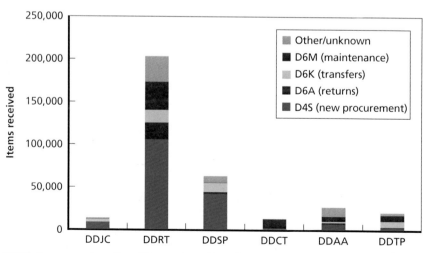

NOTE: Items are more than 50 pounds and less than 250 cubic feet.

RAND RR1375-2.2

Repair at Army Depots

When items come out of repair from Army depots, the default action is to place them in inventory at the colocated DLA depot. Item managers can adjust this process by stipulating at the time repairs are ordered that items should be stored in a different warehouse. In this case, items would be moved directly from the repair site at the Army depot to the shipping center at the colocated DLA depot, rather than to the storage section at the DLA depot. Over the period from which these data were gathered, items received at DLA depots coming from maintenance represented 17 percent of the total.

Item managers can also initiate stock-transfer orders to move items, most efficiently in truckloads, from one DLA depot to another. Items are available to issue worldwide once they are in inventory at the colocated DLA depot and frequently remain there until they are issued.

Repair at NMP Sites

Item managers may also rely on NMP sites to repair items. Through the NMP, AMC assigns maintenance for reparable items to repair sites. The NMP assesses the national-level requirement for maintenance workload and enters into agreements with approved sources of repair to perform the work. By managing maintenance centrally, the NMP ensures that work will be performed to the highest standard and reduces the redundant capability that would exist across Army posts were the work performed locally (Army Regulation 750-1, 2013).

NMP repairs may be performed at active- or reserve-component Army installations across the country. When items come out of repair at these locations, they are available for issue in the LMP. Item managers can place transportation orders to move the items to another warehouse and do so in some cases to improve responsiveness to customers, such as when there is heightened demand at a location outside the contiguous United States (OCONUS). At some NMP locations, storage space is limited, and the NMP site managers can contact item managers to request stock be moved from their site. Managers can then place a transportation order to move stock to a DLA depot.

As items are continuously being repaired, AMC can strategically position the finished items to optimize distribution performance. This is especially beneficial when item managers have less flexibility to position newly procured items whose delivery destinations are determined by a long-term purchase contract.

Figure 2.3 shows the number of NMP repairs planned for each site in FY 2015. These data include all items repaired by NMP sites, most of which weigh more than 50 pounds. The actual number of items repaired by site may varies by demand, capacity, or availability of items for repair. An additional 7,000 items were planned to be repaired at NMP sites OCONUS.

How NMP Items Are Currently Positioned

Annually, eligible sources of repair place submissions on workload they wish to perform. The NMP allocates workload for the FY, committing to fund sources of repair for workload. Throughout the course of the FY, several metrics track progress made by sources of repair. These metrics include number of repairs completed, cost of repairs, and number of unserviceable items deemed irreparable. During the FY, the NMP can change the workload it would like sources of repair to perform, subject to a set of standards determining the timeline and extent to which changes can be made.

Logistics Readiness Centers (LRCs) that participate in the NMP must store items repaired at their locations until they are sent elsewhere. LRCs repair differing quantities of items, ranging from several hundred to several thousand per year (as shown in Figure 2.3). Depending on the number of items repaired, the storage capacity of the NMP sites, and other item and LRC characteristics, repaired items can be stored at NMP sites until they are issued in response to customer requisitions. The items are available for issue in LMP.

Figure 2.3
Planned NMP Repairs, FY 2015

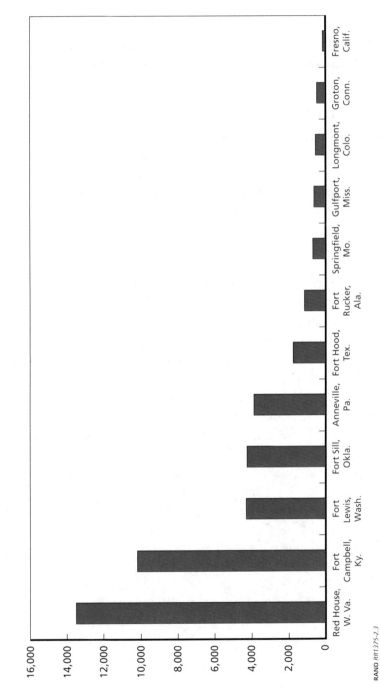

Alternatively, LRCs can send full truckloads of repaired items to DLA distribution centers.[1] In interviews with NMP staff and AMC liaisons, we learned that it can be more work for LRC staff to prepare and package large numbers of items—as individual items or small quantities directly to off-post customers—than it is to prepare and send full truckloads of items (e.g., 35–85) to DLA depots. A policy that institutes a plan for NMP sites to ship determined quantities of items to DLA hubs can decrease the labor required by NMP sites to perform their workloads. This savings will be directly recouped by NMP sites but can be passed along to AMC if NMP sites offer lower cost of labor rates to repair items.

There is no mechanism in LMP that determines when or to which DLA distribution center NMP sites should send items. Rather, LCMC item managers make this decision and determine when these transfers occur. Because item managers are rated by their ability to avoid back orders, they may choose to leave items repaired at NMP sites in place so that they are available to issue to customers directly and are not tied up in transit preparation or shipment to DLA depots. NMP staff may therefore call item managers to request permission and instructions for sending items to DLA depots, which they do to free up storage space at the LRC and to reduce workload resulting from too many individual shipments.

Recent NMP Item Stock Positioning

Figure 2.4 shows the number of items issued from each NMP site in 2013 and to what type of organization they were issued. We identify organizations that received repaired items as *local* customers located at the same installation; *DLA* depots; or other organizations, primarily Army supply-support activities at other installations (*off-post*). Most item shipments from NMP sites were to DLA rather than directly to customers. At some of the larger posts (e.g., Fort Campbell, Fort Hood, Fort Lewis), many of the items repaired in the NMP were issued to customers on-post, but even at these sites, far more items went to DLA.

---

[1]   Items sent from NMP sites to DLA depots appear as transfers in the data shown in Figure 2.2.

Figure 2.4
Maintenance Program Analysis, 2013

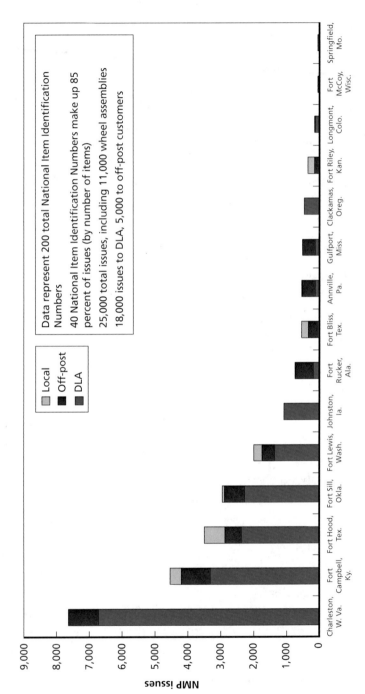

Data represent 200 total National Item Identification Numbers

40 National Item Identification Numbers make up 85 percent of issues (by number of items)

25,000 total issues, including 11,000 wheel assemblies

18,000 issues to DLA, 5,000 to off-post customers

Legend:
- Local
- Off-post
- DLA

NMP issues

Charleston, W. Va. | Fort Campbell, Ky. | Fort Hood, Tex. | Fort Sill, Okla. | Fort Lewis, Wash. | Johnston, Ia. | Fort Rucker, Ala. | Fort Bliss, Tex. | Annville, Pa. | Gulfport, Miss. | Clackamas, Oreg. | Fort Riley, Kan. | Longmont, Colo. | Fort McCoy, Wisc. | Springfield, Mo.

Because of large numbers of back orders in many AMI reparables, particularly items for wheeled vehicles, the issues from NMP sites followed a different pattern in 2015 than they did in 2013.[2] Because of the large numbers of back orders, item managers did not want to tie up inventory with stock transfers from NMP sites to DLA warehouses, so they typically issued items from NMP sites directly to customers. The different pattern of inventory management is apparent in Figure 2.5, showing that most items from the five highest-volume sites in early 2015 were issued to off-post customers rather than to DLA warehouses.

Figure 2.5
Issues from Highest-Volume NMP Sites, January–July 2015

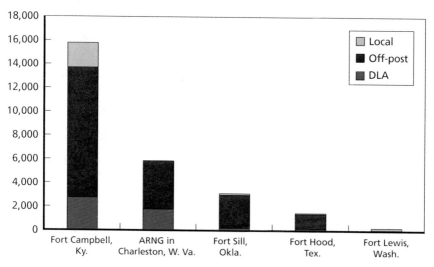

RAND RR1375-2.5

---

[2] Item managers said they predicted that the draw down in forces from Afghanistan would lead to a decrease in demand for reparables for wheeled vehicles. Given this, they adjusted inventory planning to meet a lower level of demand. However, demand for some items, such as reparables for high-mobility multipurpose wheeled vehicles, increased substantially, as these vehicles were being used more heavily during training in-garrison than they were while units were deployed. This led to large back orders for some items that took months to be filled, as item managers directed repair sites to increase production and placed new procurement orders with several-month lead times.

Item Managers Have Opportunities to Improve Stock Positioning

Strategically positioning items at DLA depots to improve distribution performance is an intuitive but complex decision that can be supported with available data. During their monthly item reviews, item managers face decisions about where to strategically position inventory. When placing orders with suppliers for new procurement items, item managers specify where suppliers should deliver them. NMP staff may seek to move truckloads of items from NMP sites to DLA depots to free storage space at their location or to minimize issuance of items to individual customers. Smaller DLA depots colocated with Army repair depots may also face storage constraints and benefit from positioning inventory at DLA distribution depots in proportion to demand.

DLA has a network of three distribution hubs at DDJC, DDRT, and DDSP. CONUS customers are divided into three regions, one for each of these centers. OCONUS customers who cannot fill demand from warehouses in theater will request items from DDJC (if they are in the U.S. Pacific Command area) or DDSP (if they are in the U.S. European Command, U.S. Africa Command, and U.S. Central Command areas). Positioning items in proportion to demands associated with DLA hubs is good practice and is the basis of the approach we recommend.

At the same time, given inventory and historic-demand data available in LMP, as well as projected transportation costs and DLA handling costs, it is possible to determine more precisely where to position stock to improve distribution performance—that is, to improve responsiveness to customers and reduce associated costs to the U.S. Department of Defense (DoD).

The overall objective should be to balance inventory across CONUS in proportion to customer needs, as well as consideration of other such characteristics as item availability, NMP site capacity, transportation speed, and distribution costs. In particular, in positioning repaired items, item managers should consider the following questions:

1.   Is there a sufficient supply to support moving?

2. Which storage locations will provide most responsive support to customers?

In this report, we discuss what affects the answers to these questions. In future research, we will provide computational algorithms to support item managers in making stock-positioning decisions, as well as numerical calculations to demonstrate how such algorithms address the objectives posed by these questions.

Discussion

For items of which there is little available serviceable inventory, item managers are concerned that the process of moving items will tie up inventory, making it unavailable to issue to customers and risking generating back orders. In such cases, it may be best to store items where they are located.

The most responsive service may be from a supporting DLA hub and the scheduled-truck service from it. Heavy items in particular may be cheaper to issue to customers via DLA depots and scheduled trucks rather than directly from more distant sites by LTL shipping. AMC may wish to maximize distribution by scheduled truck and position items at DLA depots in proportion to demand by customers. Doing so could improve responsiveness by lowering requisition wait time (RWT) and save costs to the DoD by leveraging the DLA scheduled-truck network.

In the next chapter, we will use historical data to explore responsiveness and expenses for scheduled-truck delivery.

CHAPTER THREE

Findings Related to the DLA Three-Hub Network

AMC can leverage the three-hub network that DLA has implemented in CONUS. This hub network, and the scheduled-truck service that serves it, enables DLA to distribute items to customers quickly on a predictable schedule and at reasonable cost. DLA pays for dedicated trucks to deliver items from its primary distribution centers to its highest-volume customers, departing on a consistent weekly schedule. Below, we discuss how this network works and how it may serve Army needs as well.

Shipments by Scheduled Truck Are Faster than by Other Modes

For items that are positioned at DLA hubs, distribution by scheduled-truck service can be faster than by commercial truck shipment from other sites. Figure 3.1 shows RWT for items shipped by scheduled truck and by commercial truckload or LTL shipments, as well as the average difference between them.[1]

For most locations, scheduled-truck shipments are delivered to customers, on average, five days faster than are LTL or truckload shipments by commercial carriers.

---

[1] Some installations may get scheduled-truck shipments from more than one DLA depot. For our analysis, we consider only scheduled-truck shipments from the DLA depot best aligned with an installation.

Figure 3.1
Average Requisition Wait Time from DLA Depots to Customers, January 2014–July 2015

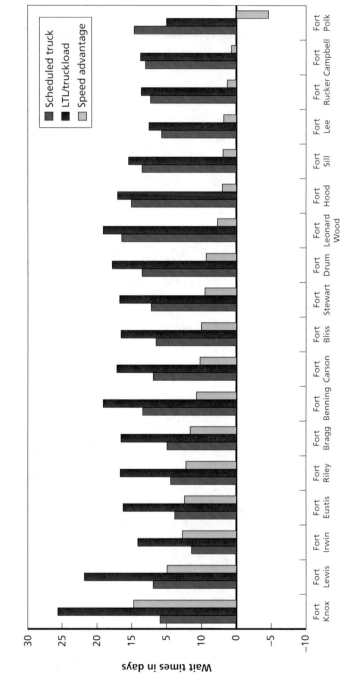

The lengthy LTL and truckload RWTs can be attributed to a number of causes. LTL and truckload shipments may originate from a DLA depot that is distant from a customer and that issues items to the customer infrequently. When an LTL and truckload shipment is initiated, an order is placed with a commercial carrier. The carrier will send a truck to pick up the item when it is economically efficient for the carrier to do so. The truck may be partially full of items from other customers and not have space to carry the entire set of items ordered to be shipped, so part of a shipment may wait until the carrier sends another truck. When items are picked up, the carrier's truck will transport the items from the DLA depot to the carrier's distribution center, where the items will wait until another truck is available to carry them along the next leg, either to another distribution center or the Army customer. All of these steps take time and can add up to lengthy RWTs.

Alternatively, when items are shipped via scheduled truck, they are picked up at scheduled times—up to four times per week. Items are carried by a dedicated truck along a direct route, although the route may contain several DoD customers. After pickup, all items are delivered within several days.

Fort Lewis is a case where the RWT for scheduled-truck shipments is much faster than for LTL and truckload. Fort Lewis receives three trucks per week from DDJC. Fort Lewis is in Oregon, and DDJC is in Northern California, so they are relatively proximate. When Fort Lewis receives items via LTL or truckload, they often originate at DDRT or DDSP, which are a much greater distance away.

Fort Polk is an example of a customer that is not well served by the scheduled-truck network. Fort Polk receives one truck per week from DDSP. Fort Polk is in Louisiana, and DDSP is in Pennsylvania, so the distance is fairly great. When Fort Polk receives items via LTL or truckload, the items often originate from DDRT, which is a short distance away in eastern Texas.

Fort Polk is scheduled to receive one scheduled truck per week from DDSP on the basis of its demands. The vast majority of items Fort Polk demands are managed by DLA rather than the Army. DLA continues to store most of its inventory at DDSP and little at DDRT. If DLA positions more inventory at DDRT, it could improve responsiveness to all customers in the central region, in addition to nearby Fort Polk.[2] However, DLA enjoys warehousing efficiencies by locating inventory primarily at one location, DDSP, and may be slow to increase inventory at DDRT.

Opportunity to Save Costs Leveraging the DLA Scheduled-Truck Network

DLA directly bears the cost of issuing items from its warehouses to CONUS customers. It recoups these costs through the cost-recovery rate it imposes on items that are sold from the Defense Working Capital Fund to customers from the military services and from net landed costs it assesses to items received, stored, and issued from its warehouses. DLA established its scheduled-truck network in coordination with the military services to garner savings on shipping costs and improve responsiveness. The Army can leverage this scheduled-truck network to save costs to DoD of shipping items from storage warehouses to customers. By doing so, the Army will also improve responsiveness to its customers.

DLA contracts trucks to depart its warehouses, mainly its hubs, to carry items to customers. The trucks may depart a warehouse and travel to a single customer location, or they may deliver items to two or three military customers near one another. Depending on the volume of items going to customers, the trucks may be scheduled to depart one to five times per week. The median delivery schedule is about two times per week.

The trucks are, on average, one-half to two-thirds full. This leaves room to add materiel to the delivery routes at low cost. Were the AMC to position stock to significantly increase the volume of items carried on the truck routes, then DLA could increase the frequency of the routes. While this would increase costs, it would also improve respon-

---

2   Fort Polk and Army customers in Louisiana are in the eastern region, as designated in LMP. The dividing line between eastern and central regions is the Mississippi River. Presently, this alignment is practical for Fort Polk, as it receives one scheduled truck per week from DDSP. Other Louisiana customers, such as ARNG units, may benefit from being aligned with DDRT and the central region in LMP. If items they request are available at DDRT, LTL shipments may be a few days faster.

siveness and decrease RWT. Because of the great benefit of leveraging the scheduled-truck routes to issue items to customers, and, as we discuss below, the low cost relative to other modes of issuing items to customers, we attach no cost to scheduled-truck transportation for the purposes of this analysis.[3]

The benefit of saving transportation costs is greatest for heavier items, as shipping rates are structured on a per-pound, per-mile basis. Depending on the distance, it costs approximately between $0.25 and $0.75 per pound to issue items to customers by LTL shipping (as determined from regression equations of data on several thousand LTL shipments from DLA warehouses to customers).

Our analysis, as noted earlier, focuses on positioning AMI that weighs more than 50 pounds but is less than 250 cubic feet in volume. Such items typically include class IX repair parts for vehicles or aircraft. The most common items in this weight class, by volume of items issued, are wheel assemblies, track shoes, engines, and transmissions.

Figure 3.2 shows the portion of AMI issued that are in this weight class among the six DLA depots that handle the most AMI. The items in this weight class make up 6 percent of items issued but 81 percent of the total weight.

Table 3.1 shows the costs of several shipments that will be used to compare options. To consider the cost of shipping an item via LTL from a warehouse to a customer, we approximate this as $150. We approximate the cost to transfer an item via truckload to a DLA depot, then to handle the item at the DLA depot, with a cost of $81. Finally, we assume no marginal cost, or a cost of zero dollars to ship an item

---

[3] If increased volume on DLA scheduled trucks does not cause a necessary increase in truck volume, then there will be low additional transportation costs because of an increase in strategic stock positioning. In cases where the increased volume does lead to additional scheduled trucks transporting items from DLA depots to some customers, then there will be an associated cost with these shipments; however, there will be an even greater improvement in responsiveness and decrease in RWT than predicted in Figure 3.5. Were this to occur, it could be reasonable to assume that the cost per pound of transporting items by DLA scheduled truck would be approximately $0.20 per pound. While the cost savings predicted in Figure 3.5 decrease from approximately $9 million per year to $6.5 million per year, the decrease in RWT and improved responsiveness would be even greater than predicted in the chart.

Figure 3.2
AMI Issued from DLA Warehouses

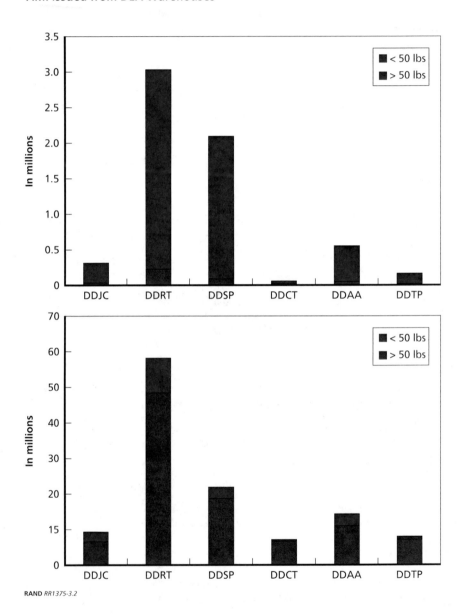

Table 3.1
Costs of Shipments from Different Locations and by Different Modes

Estimated Cost to Ship 250-Pound Item	
Ship via LTL to customer at $0.60 per pound	$150
Transfer via truckload ($0.20 per pound) to DLA depot, receipt and issue by DLA	$81
Ship via scheduled truck to customer	n/a

via scheduled truck from a DLA depot to a customer on a scheduled-truck route.

Consider the example in Table 3.1, where a 250-pound wheel assembly is issued from a warehouse and transported by LTL shipping through such a great distance as from DDRT to Fort Lewis, at a cost of $0.60 per pound, for a total cost of $150 to transport the item. If an item manager directs a supplier to deliver items to DLA warehouses in proportion to demand, then, for each item that is issued via scheduled truck rather than LTL from a distant warehouse, DoD will save approximately $150. The savings would directly accrue to DLA but later passed on to the military services by the cost-recovery rate and net landed cost assessments.[4]

Where items are already in inventory and stock is not balanced across warehouses, it can also be cost-effective to transfer items to leverage the DLA scheduled-truck network. The cost to transport an item across the country as part of a full truckload of items, then receive it and issue it from a DLA warehouse, is approximately $81, as shown in Table 3.1. So, shipping such items by such a transfer to a DLA hub then by scheduled truck from the hub to the customer would yield a net savings to DoD of $69. The most likely instance where stock is transferred between warehouses may be when there are capacity constraints

---

[4] The shipping distance from DDRT to Fort Lewis was chosen as a representative example. Shipments from DDRT to Fort Lewis (in Washington) and Fort Irwin (in California) represent a large portion of the out-of-region LTL shipments for the items in this study. The shipping costs for this example are in the middle range of out-of-region LTL shipments, higher than shipping costs from DDSP to Fort Lewis and Fort Irwin, but lower than shipping costs from DDRT to Fort Drum (in New York).

forcing the move. Some of the smaller DLA depots, particularly those colocated with repair sites, may, as noted earlier, have space constraints. NMP sites may also have space constraints and wish to move repaired items to another site, such as a DLA depot.

Item managers presently cannot use LMP data so as to position stock at DLA depots in proportion to demand. We recommend that such a feature be developed.

Business Case for AMC Stock Positioning

AMC can improve responsiveness to customers by positioning items at DLA depots, which, in turn, can issue items by scheduled-truck service. This option can reduce RWT by five to ten days relative to LTL shipments from more distant warehouses. DoD will save distribution costs, directly accrued by decreased transportation costs paid by DLA, when it issues items to customers in CONUS. These savings can later be passed on to the military services when they reimburse DLA.

By focusing on secondary items weighing more than 50 pounds, AMC can strategically position items that represent 81 percent of the total weight of AMI secondary items issued from DLA depots. Such items make up roughly 400,000 of the 4.2 million AMI secondary items issued by DLA depots from January 2014 to May 2015. These approximate 400,000 items can be identified in Figure 3.3 by the data point labeled "Issues from Depot to All Customers."

Of these 400,000 items, approximately 180,000 were issued to customers who were also on CONUS scheduled-truck routes. Issues to scheduled-truck customers can be identified in Figure 3.3 by the data labeled "Issues from All Depots to Scheduled-Truck Customers in Region." For purposes of this chart, all CONUS Army customers who are on scheduled-truck routes are distinguished as being in the western DDJC region, the central DDRT region, or the eastern DDSP region. Since they are colocated, the Army depot in Corpus Christi was defined as being a scheduled-truck customer of DDCT. The Army depots at Anniston and Tobyhanna were similarly defined as being scheduled-truck customers of DDAA and DDTP, although, in these cases, the method to items from DLA depot to Army depot is by local truck or forklift, rather than a scheduled truck.

Figure 3.3
Issues of AMI Greater than 50 Pounds, January 2014–May 2015

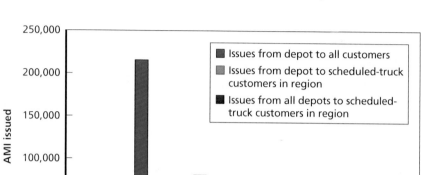

RAND *RR1375-3.3*

Half of these 180,000 items were aligned issues, items issued from DLA depots to customers on their scheduled-truck routes. These approximate 90,000 issues can be identified in Figure 3.3 by the data point labeled "Issues from Depot to Scheduled-Truck Customers in Region." The other half were issued from more-distant DLA depots and shipped by other modes; there were about 90,000 items that could have been delivered by scheduled-truck service if they were strategically positioned—then RWT could have been decreased and responsiveness to Army customers could have improved.

Many secondary items issued by DLA are new procurements. The data in Figure 3.4 show that, over the same period, 50 percent of AMI received at DLA depots were acquired by this method. An additional 11 percent of items were received at DLA depots as transfers, many of these receipts were items transferred from NMP sites. In both of these cases, item managers can make stock-positioning decisions to improve distribution performance with potentially little increased cost.

If AMC item managers were able to position inventory proportionate to demand at the time they initiated new procurements and

Figure 3.4
Receipts of Serviceable Items at Highest-Volume Army Depots in the United States

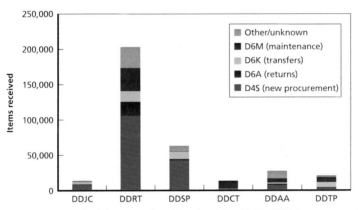

NOTE: Items are more than 50 pounds and less than 250 cubic feet. This figure repeats data in Figure 2.2.
RAND *RR1375-3.4*

transferred items from NMP sites, they could reduce average RWT for the affected items and save distribution costs by several hundred thousand dollars per month. The data in Figure 3.5 also show benefits to DoD if AMC item managers considered distribution performance when positioning all items in this group. A full implementation would include strategic positioning of customer returns and excess, as well as transferring items that come out of maintenance at Army depots when it is advantageous.

Were all of the items in this analysis delivered by scheduled truck from DLA hub rather than by LTL shipping, 90,000 items would be affected, and DoD would have saved nearly $13 million over the 17-month study period, which translates to more than $9 million per year. Were AMC to undertake an initiative to improve stock positioning, responsiveness would improve, and costs would be saved by distribution cost savings.[5]

---

[5] This calculation represents the total number of affected items—90,000—multiplied by the estimated $150 cost savings to ship a 250-pound item via scheduled truck rather than LTL. The average weight of an item in this study is 250 pounds.

Figure 3.5
Savings in Requisition Wait Time and Distribution Costs Because of Stock Positioning; AMI Issued to Scheduled-Truck Customers from Distant DLA Hubs

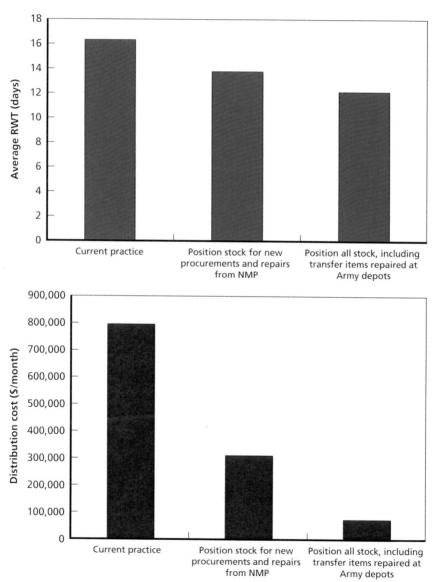

In summary, when it is necessary to move items for such reasons as freeing up storage space at repair sites, item managers have the opportunity to improve responsiveness and save costs by moving items to DLA hubs in proportion to demand in each region. This will enable the Army to leverage the DLA scheduled-truck network, improve responsiveness, and reduce transportation costs. DLA would also favor this strategy, as it would support AMC item manager decisions if they are asked to move items from smaller DLA depots (such as DDCT) that can become space constrained to larger DLA depots (such as DDRT and DDJC) that have much greater capacity.

By strategically considering placement of items before shipment, item managers can improve responsiveness and save costs. If AMC leverages the capabilities of LMP to practice strategic stock positioning, it would change its management principles to differ from current practice in ways that are summarized in Table 3.2.

In the next chapter, we review a case study of one particular item for which AMC has already sought to improve responsiveness.

Table 3.2
Management Principles Under Current Practice and Strategic Stock Positioning

Current Practice	Strategic Stock Positioning
No visibility of regional demand	Transaction code in LMP calculates regional demand
No decision support to position stock	Transaction code in LMP compares demand projection with inventory projection
Item managers place new procurements orders to single warehouse	Item managers can align orders regionally, improving RWT and lowering distribution costs
Repair sites must contact item managers to move items, leads to excess and misaligned inventory	Item managers have decision support to tell repair sites in advance where they should ship completed items
Item managers do not have a business case to move items that come out of repair at Army depots	Item managers can justify moving items to leverage scheduled-truck network, decrease RWT

CHAPTER FOUR

AMC Successful Implementations

Success of AMC Stock Positioning of Wheel Assemblies

In the initial stages of this research, RAND Arroyo Center and AMC performed a case study of wheel assemblies to learn how NMP items are managed and to identify implementable recommendations. Wheel assemblies are a notable item to consider for distribution efficiencies, as they generate a high quantity of AMI repairs. As shown in Figures 4.1 and 4.2, the WV ARNG in Charleston, West Virginia, repaired about 60 percent of wheel assemblies in the NMP in 2012 and 2013.

All of the wheel assemblies the WV ARNG sent to DLA depots in 2012 went to DDRT. Item managers reported that, during the height of operations in Iraq and Afghanistan, DDSP experienced delays distributing materiel because of overwhelming demands placed on it as DoD's primary distribution site on the East Coast. Item managers found they could improve distribution speed by sending items to DDRT for further distribution through the Gulf of Mexico and onto Iraq and Afghanistan via surface freight. While these were sensible reasons for this decision, sending all these wheel assemblies to DDRT was not the most-efficient option to meet CONUS demand.

In March 2013, we recommended that AMC direct TACOM to have the WV ARNG send some of the wheel assemblies it ships to DLA to DDSP rather than to DDRT. AMC G-3/4 issued a memo instructing item managers to direct WV ARNG to begin sending items to both DDRT and DDSP. Figure 4.3 illustrates the change of trends by month in the number of wheel assemblies issued to the three DLA depots.

Figure 4.1
Total Wheel Assemblies Repaired, 2012

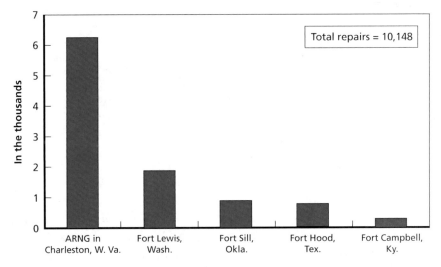

RAND RR1375-4.1

Figure 4.2
Total Wheel Assemblies Repaired, 2013

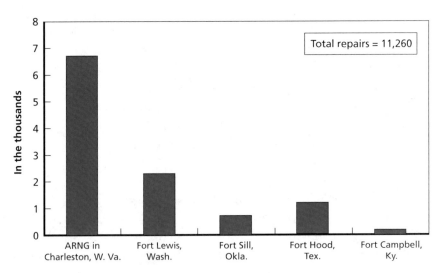

RAND RR1375-4.2

Figure 4.3
WV ARNG Wheel Assembly Issues, February 2012–August 2014

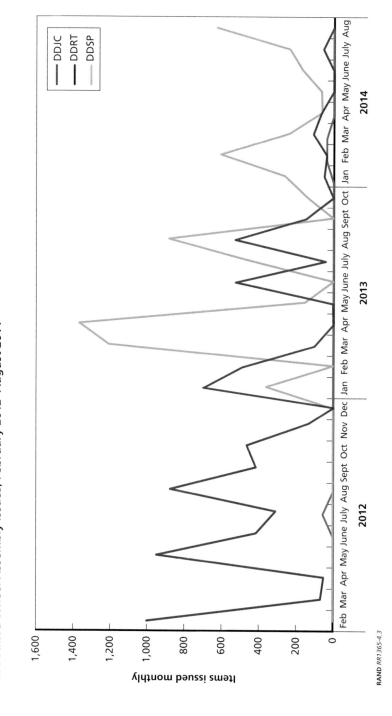

RAND RR1365-4.3

This change enabled Army customers on scheduled-truck routes from DDSP to receive wheel assemblies via this mode in 2013. Figure 4.4 shows the proportion of DLA wheel assembly issues from each DLA hub to customers in the eastern region of the United States from 2012 to 2014—and the increasing proportion that came from the DDSP for these customers.

In terms of the same metrics used to project savings to the Army by implementing stock-positioning policies for all AMI, we estimate the effect of change in wheel assembly management. As shown in Figure 4.3, the national maintenance program at WV ARNG sent approximately 4,200 wheel assemblies to DDSP for later distribution to customers and 500 wheel assemblies to DDRT in 2014. We can estimate that 23 percent of the wheel assemblies WV ARNG sent to DDSP were eventually issued to customers on its scheduled-truck

Figure 4.4
Wheel Assembly Issues from DLA Hubs to Eastern Region Customers

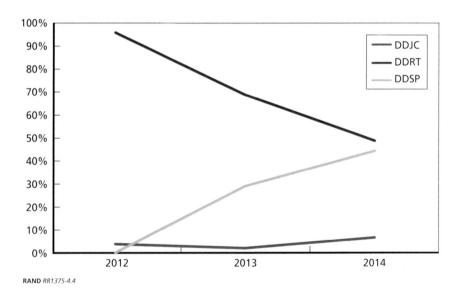

RAND RR1375-4.4

routes.[1] Based on the estimated number of additional wheel assemblies distributed to customers via scheduled-truck routes, the projected savings of $150 to issue a wheel assembly to a customer via scheduled truck rather than LTL (Table 3.1), and historical distribution times observed for use of scheduled truck and LTL, we estimate the Army saved distribution costs of approximately $150,000, and reduced RWT for the affect wheel assembly issues by one day (Figure 4.5). We estimate that distribution costs decreased by approximately $150,000, and RWT was reduced by one day.

AMC can continue to improve the support its customers receive by ensuring that items positioned at DLA depots are delivered to the intended customers. Figure 4.6 shows DLA depots frequently issued NMP items to customers outside their CONUS regions, as well as to customers within their region but not on scheduled-truck routes. Such shipments can be more costly than those from DLA depots within region or from NMP sites directly. When items positioned at DLA depots are issued to customers not on scheduled-truck routes, the inventory at DLA depots is depleted and unavailable to meet requests from customers who could benefit from receiving items via DLA scheduled truck.

Through the capabilities of LMP, AMC can tailor the source-preference logic that selects the warehouse for issuing an item to a customer. Below, we describe the LMP source-preference logic and the AMC pilot to use it to tailor how warehouses are selected to issue items to customers.

AMC Pilot Implementation of LMP Source-Preference Logic

LMP is the Army ERP system. It receives requisitions for AMI and processes them to issue items to customers either automatically or with the input of an item manager. To automatically process issues, the system refers to a set of files that manage the source-preference logic. The files

---

[1] DDSP receives serviceable wheel assemblies from multiple sources in addition to WV ARNG, including suppliers sending new procurement items. We calculated the portion of each type of wheel assembly that was issued from DDSP to scheduled-truck customers in 2014, and used these portions to calculate a weighted average of the wheel assemblies sent from WV ARNG to DDSP. The weighted average is 23 percent.

Figure 4.5
Estimated Effect of Wheel Assembly Positioning Policies, WV ARNG Issues to DLA; Reduced Customer RWT and Distribution Costs in 2014

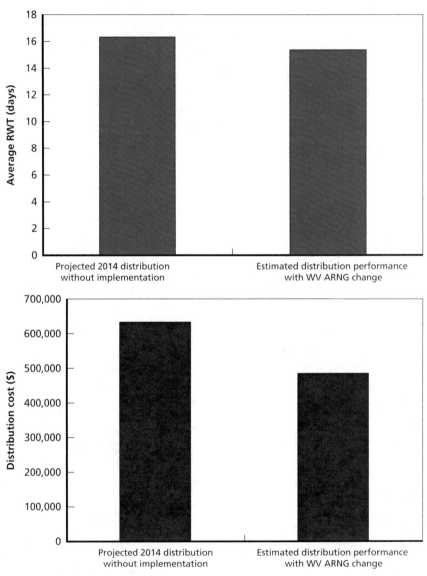

Figure 4.6
Customers to Whom NMP Items Were Issued from Each DLA Depot

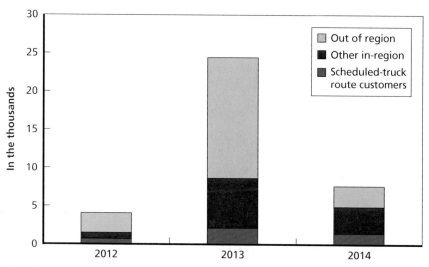

RAND RR1375-4.6

are called the Routing Identifier Code/Department of Defense Activity Address Code (RIC DODAAC) file and the ZSDDEPOTS file (a title of a file/table used in LMP). The first file is standard for all AMI, and a version of the second file is managed by each of the three LCMCs: U.S. Army Aviation and Missile Command, U.S. Army Communications–Electronics Command, and TACOM. These files are used in a set of three searches performed to identify inventory at warehouses that can be issued to customers. The first search is called the *on-installation search*. The second is called the *national-level search*. The third search is called the *off-installation search*. These searches are performed in order: As soon as a warehouse is identified with an item that can be issued, LMP initiates the issue.

The on-installation search contains warehouses local to the requesting customer. The national-level search includes DLA warehouses and then NMP sites. The third search includes warehouses at other Army installations that are not generally used as distribution depots. Within each of the searches, a list of warehouses is checked in an order that generally begins close to a requesting customer and proceeds geographically outward.

Most stock available to issue is held at DLA warehouses and NMP sites. So it is within the national-level search that many requisitions are filled. The priority in this case is to issue items from DLA warehouses if available there and then check NMP sites if necessary for inventory. In some instances, this may not be a preferable set of priorities.

When NMP sites repair items and then ship them as full truckloads to DLA depots for distribution, the most-efficient way to get those items to customers is to use the DLA scheduled-truck routes. To improve distribution performance, as many items as possible should be issued by this mode. But item requests from both customers on scheduled-truck routes and other customers are routed identically to be filled from DLA warehouses. Presently, there is no way to distinguish customers by whether they can be served by DLA scheduled trucks.

This research identified an opportunity to use the on-installation search as a way to promote NMP sites as an item source and to issue an item to customers preferentially over DLA warehouses. Since the on-installation search is performed before the national-level search, customers who have NMP sites added to their on-installation search will be issued items from such sites when available. As the on-installation search is defined for each customer, it is possible to add NMP sites to the on-installation search for selected customers. Issuing items from NMP sites directly to customers not on scheduled-truck routes will have little effect on responsiveness, as these customers receive items by LTL shipment regardless of the originating location. Satisfying demand for customers not on scheduled-truck routes from NMP sites will preserve inventory at DLA depots so that it is available to customers on scheduled-truck routes. There is precedent for using the on-installation search to prioritize warehouse selection. The Army placed the warehouses at Sierra Army Depot in the on-installation search for all CONUS Army customers to strategically deplete inventory from that location. In Europe, Army customers have warehouses in Germersheim, Germany, in their on-installation search, for similar strategic reasons.

The NMP site at the WV ARNG (whose RIC is W55) is the best place to pilot this strategy, as the only items in inventory at that location are items repaired as part of the NMP. So including W55 in the on-installation search will result in only items repaired as part of the NMP being issued to customers. By contrast, at other NMP sites, such as Fort Campbell or Fort Hood, where thousands of various items are recorded in inventory, nearly all of these items are either to be consumed by the repair operation at those sites or are intended to be issued to customers at those installations. Issuing them in response to national-level customer demand would be a very disadvantageous outcome, resulting in shortages at those installations and increasing distribution costs. Further research can identify ways to distinguish which items at NMP sites are intended for national distribution and which would enable the pilot implementation performed at W55 to be rolled out across NMP sites.

The pilot implementation to include W55 in the on-installation search for eastern region customers began on May 28, 2015. The intent was to adjust the source-preference logic for filling requisitions in LMP so that smaller customers not served by DLA scheduled trucks would be issued items preferentially by the WV ARNG rather than an eastern region DLA depot. Two of the largest nonscheduled-truck route customers in the eastern region are the Tennessee and South Carolina ARNG units. The many locations of ARNG units in these states order items independently, but their requests are routed through central U.S. Property and Fiscal Officer (USPFO) sites, of which there is one in each state. The Supply Support Activity at the USPFO passes requisitions to the LMP, but the requisition contains information about the originating customer, so when a warehouse issues an item to a customer, the item can be sent directly to the customer rather than to the USPFO site. The two RICs for the Tennessee and South Carolina ARNG units are A61 and A62.

On average, ARNG units from these two states will collectively be issued 20 to 25 items per month from the NMP site at the WV ARNG. During the two months that the pilot implementation was in place, they were issued approximately 250 items from the WV ARNG NMP site. They were issued zero items of these types from DLA during this period, although DLA had available inventory. This result shows the pilot implementation worked: It had the intended effect of preferentially issuing items from W55 for these customers before issuing items from DLA depots.

By issuing 250 items to these customers over the pilot implementation, the set of requisitions passed to DLA changed in its makeup. The proportion of demands from customers on scheduled-truck routes increased. Based on the mix of customer requests filled by DDSP during this period, we estimate that DLA issued 40 more items to scheduled-truck customers than it would have were the pilot not in place. However, the net amount of items issued by DDSP and W55 should not have changed and was observed to be similar to previous months.

The effects of this pilot may have been muted because of the low balance of wheel assembly inventory across DLA warehouses and NMP sites during the period. This led to imbalanced inventory, and the warehouses in the eastern region—including DDSP and W55—issued more items to customers outside the region than they had historically. Had they been issuing a larger portion of items to customers on DDSP scheduled-truck routes, the effect of the implementation could have been greater, resulting in a greater increase in issues to scheduled-truck customers.

Future Implementations

RAND developed a spreadsheet that can process LMP data to generate reports that aid item managers in making stock-positioning decisions. By sharing this spreadsheet with managers responsible for items of the type analyzed in this research, AMC can begin to appreciate the advantages presented in the business case.

To make stock positioning a long-term strategy, the capability present in the spreadsheet can be adapted to become a LMP transaction code that can be run by item managers and supply-chain management staff across the Army and AMC.

AMC can direct its attention to positioning stock at DDJC and DDSP, as they represent the greatest opportunities to improve distribution performance. Figure 3.3 shows the quantity of AMI greater than

50 pounds that were consumed by aligned customers—that is, customers on DLA-depot scheduled-truck routes (in the dark-red column) and the quantity of that consumption that was met by the preferred DLA depot (shown in the light-red column). It also shows, in the blue column showing total issues, that the depots have sufficient capacity to issue the items to meet their aligned customers' needs.

Figure 3.4, which shows how items arrived at DLA depots, helps indicate opportunities to improve stock positioning. Many of the items that are received at DDRT are new procurements, while there are few new procurements at DDJC. One of the greatest opportunities to improve distribution performance, based on historical consumption, would be to balance new procurement across the DLA hubs. A key part of implementing this strategy will be analyzing historical consumption to assess whether it is likely future consumption will be similar.

CHAPTER FIVE

Implications and Recommendations

AMC can improve distribution performance through stock positioning. The two main opportunities to direct where inventory is located are when:

1. items are newly procured
2. items come out of maintenance.

New procurement is the main method by which items are made available to issue to customers, and it presents the greatest opportunity to improve distribution performance through stock positioning. By selecting the destination where newly procured items are delivered so inventory at distribution hubs is proportionate to demand, AMC can greatly improve distribution performance for Army-managed items.

For secondary items, positioning heavy items at distribution hubs to leverage the DLA scheduled-truck network can result in both increased responsiveness to customers and decreased distribution costs to DoD.

When items come out of repair at Army depots, they are typically placed into inventory at colocated DLA distribution hubs. While it can generate cost savings to DoD to transfer heavy items to balance inventory across hubs proportionate to demand and increase use of the DLA scheduled-truck network, doing so carries an upfront cost to the Army and will likely only be a best option when customer service is highly impinged. The upfront cost is a small trade-off to the Army for the benefit of reducing wait times for its customers.

In the case where items come out of repair at NMP sites, frequently the NMP sites request that AMC move the items to DLA distribution hubs to relieve capacity constraints at NMP locations. When this occurs, AMC can make the strategic decision to position inventory at DLA hubs proportionate to demand to improve distribution performance.

Currently, data in LMP are not readily available to support item managers in making stock-positioning decisions to improve distribution performance. By working with item managers, AMC can identify the best format for this information to be generated. RAND Arroyo Center has developed a prototype spreadsheet tool that uses LMP output to create reports to support stock-positioning decisions. Using this tool can be a constructive way to demonstrate the ease and benefit of using LMP data to make stock-positioning decisions when initiating purchases of new items and positioning items that come out of repair. By having information to make stock-positioning decisions that balance inventory against regional demand, item managers can save costs and lower RWT.

References

Army Regulation 750–1, "Maintenance of Supplies and Equipment Army Materiel Maintenance Policy," Washington, D.C.: Headquarters Department of the Army, September 12, 2013.

Dumond, John, Marygail K. Brauner, Rick Eden, John R. Folkeson, Kenneth J. Girardini, Donna J. Keyser, Eric Peltz, Ellen M. Pint, and Mark Y. D. Wang, *Velocity Management: The Business Paradigm That Has Transformed U.S. Army Logistics*, Santa Monica, Calif.: RAND Corporation, MR-1108-A, 2001. As of June 1, 2014:
http://www.rand.org/pubs/monograph_reports/MR1108

Saliba, Gabriel, "Beyond Sustainment: How the U.S. Army Logistics Modernization Program Is Taking Materiel Delivery and Accountability to the Next Level," *Army AL&T,* January–March 2013, pp. 78–83. As of November 5, 2014:
http://asc.army.mil/web/wp-content/uploads/2013/04/Jan-Mar2013_army_al.pdf